AI AWARENESS SERIES

AI Governance
and Ethics

Andrea V.M. Greaves

Contents

Introduction

Welcome to AI Governance and Ethics, part of our AI awareness training series. This book introduces the key ideas shaping how AI is managed, regulated and applied responsibly in today's world as AI technology moves fast, questions of trust, fairness and accountability are more important than ever.

You'll explore how governments, companies and global institutions are working to set the rules and build public confidence. We'll look at emerging trends in AI policy, the ethical risks of automation and bias, and the frameworks being developed to ensure AI systems are transparent and explainable.

You'll also consider how topics like data privacy, social inclusion and environmental responsibility connect to AI decision making, from the role of blockchain in building trust to efforts around equity and diversity in tech. This book offers a broad foundation in what ethical AI can look like in practice.. Let's get started.

Chapter 1: Introduction to AI Governance

AI governance refers to the set of policies, standards, and frameworks designed to ensure that artificial intelligence is developed and used responsibly. It's about making sure that AI serves society's best interests and operates in a safe and ethical way. Importantly, governance applies across various sectors and AI applications, defining the extent of oversight needed.

At its core, AI governance provides a framework for guiding how AI is developed and used. The goal is to promote innovation while also addressing key concerns like ethics, law, and societal impact. It's about balancing progress with responsibility.

AI governance frameworks include broad ethical guidelines as well as specific regulations that guide both the creation and deployment of AI. But it's not straightforward. AI's rapid development and technical complexity make it difficult to regulate, and different countries have varying laws, which complicates enforcement.

Chapter 1: Introduction to AI Governance

Let's look at some practical examples of AI governance in action. Data protection laws help regulate how personal information is collected and used in AI systems. AI ethics boards within companies oversee responsible AI practices. On the global stage, international agreements help establish standards for AI development and encourage collaboration between countries.

Transparency is key to building trust in AI systems. When AI operations are clear and understandable, users are more likely to trust them. Transparency also helps regulators oversee AI activities to make sure they're operating fairly and responsibly.

Accountability means that AI developers, deployers, and users are held responsible for the outcomes of AI systems. Mechanisms must be in place to ensure ethical behavior and provide ways to address any harm caused by AI. It's about making sure no one can shirk responsibility when AI causes an issue.

Fairness is about ensuring AI systems don't discriminate or reinforce biases. Developers must work actively to mitigate bias and ensure that AI treats all users equitably. This helps build inclusive systems and fosters public trust.

Chapter 1: Introduction to AI Governance

Governments play a crucial role in AI governance. They create laws and set standards that ensure AI technologies are safe and ethical. Regulators then enforce compliance with these laws, all while balancing innovation with the protection of public interests like privacy and fairness.

Corporations are responsible for developing and deploying AI, often leading technological innovation. Academia contributes by researching AI impacts and ethics, helping to guide responsible development. Civil society organizations act as watchdogs, advocating for accountability, inclusivity, and ethical standards.

Chapter 1: Introduction to AI Governance

Effective AI governance relies on collaboration across sectors—governments, industry, academia, and civil society all need to work together. Sharing knowledge and best practices strengthens governance. And by aligning their goals, these stakeholders can ensure AI is developed and governed in a responsible, ethical way.

Different countries have different societal values, which affect how they approach AI governance. Their legal systems also shape regulatory frameworks, and economic priorities can influence how nations invest in and control AI. These variations make global governance a complex challenge.

Chapter 1: Introduction to AI Governance

When it comes to cross-border AI regulation, conflicting laws can create legal grey areas. Sometimes, no clear jurisdiction exists to enforce rules effectively. And even when laws are in place, enforcement can be tricky due to differing national priorities and limited cooperation between countries.

Despite the challenges, international cooperation presents real opportunities. By working together, countries can harmonize regulations, share best practices, and address shared risks. Global collaboration is key to responsible AI governance on a worldwide scale.

Chapter 1: Introduction to AI Governance

In conclusion, AI governance is about ensuring artificial intelligence benefits society while minimizing risks. It requires a careful balance between fostering innovation and enforcing ethical standards. With the right principles, stakeholder collaboration, and international cooperation, AI can be governed responsibly for the good of all.

Chapter 2: Ethical Frameworks for AI

Ethical considerations are essential in AI development for several reasons. First, AI must always respect fundamental human values and rights. Second, ethical AI aims to prevent harm—whether physical, emotional, or social. Third, transparency and accountability are critical for building public trust. And finally, integrating ethics early in the design process helps prevent unintended negative consequences and supports responsible innovation.

Let's look at some of the primary ethical frameworks applied to AI: Deontological ethics focuses on duties and rules, emphasizing moral obligations regardless of outcomes. Utilitarianism evaluates actions based on the outcomes they produce, aiming for the greatest happiness. Virtue ethics centers on the character traits AI should promote, like honesty and fairness. Care ethics highlights empathy, guiding AI to act with care and responsibility in its interactions.

Chapter 2: Ethical Frameworks for AI

Applying ethical principles to AI isn't always straightforward. Ambiguous accountability arises because rapid AI advances blur lines of responsibility. Evolving societal norms make it challenging to keep AI ethics aligned with changing values. Unpredictable impacts mean we can't always foresee the long-term effects of AI. Therefore, ethical frameworks must adapt continuously as technology evolves.

Utilitarianism is about evaluating actions by how much they increase overall happiness or utility. Consequentialism similarly focuses on the outcomes of actions to determine their morality. Both are highly relevant to AI because they provide a way to assess whether AI actions are promoting well-being and positive societal impact. To apply these principles, AI decisions should be assessed based on their impact on all stakeholders. The aim is to maximize positive outcomes, promoting actions that benefit the majority. At the same time, minimizing harm is crucial—AI systems must strive to reduce negative consequences. This is the core of the utilitarian approach: doing the greatest good for the greatest number.

Let's look at examples of utilitarian reasoning in AI: In healthcare, AI may prioritize treatment allocation to maximize benefits across the patient population. In autonomous vehicles, AI might make split-second decisions to minimize harm to both passengers and pedestrians. These scenarios show how utilitarian ethics guide AI decision-making in real-world applications.

Rights-based ethical theories emphasize the importance of human dignity and autonomy in AI systems. Empowering users means giving people control over how AI affects them. Respecting personal dignity requires AI to operate without bias or discrimination. Protecting privacy ensures that AI doesn't misuse personal data or violate confidentiality. Balancing user rights with technological progress is key. Developers must innovate responsibly, ensuring new technologies don't compromise user privacy or ethical standards. Protecting user rights means safeguarding transparency, fairness, and autonomy even as AI advances.

Chapter 2: Ethical Frameworks for AI

When we talk about justice and equity in AI: Justice means fair treatment and balanced distribution of AI's benefits and harms. Equity is about recognizing and correcting systemic inequalities that AI could worsen. Both concepts aim to ensure AI supports fairness in society.

Algorithmic bias is a major ethical concern. Bias can stem from flawed data or design choices. Identifying bias early is critical for fairness. Mitigating bias means implementing strategies to prevent discrimination and ensure equitable outcomes from AI systems. Promoting fairness in AI decision-making involves several strategies: Using fairness-aware algorithms designed to minimize bias. Ensuring transparent processes, so users understand how decisions are made. These approaches help foster trust and ensure ethical AI behavior.

Autonomous systems bring unique ethical challenges. They make decisions without direct human oversight, which raises concerns. They may encounter moral dilemmas with conflicting values. And their actions can lead to unpredictable outcomes—so careful ethical consideration is needed. Responsibility in autonomous systems is complex. When harm occurs, assigning responsibility isn't always clear. That's why we need clear accountability frameworks. Legal systems also have to evolve to address liability issues arising from AI decisions.

In real-world deployment, ethical considerations are vital. Safety assurance means AI should never put people at risk. Transparency builds user trust by explaining how AI makes decisions. And assessing societal impact ensures AI benefits communities and adheres to ethical standards.

To wrap up, ethical frameworks are essential in guiding responsible AI development. By considering human values, fairness, accountability, and the broader social impact, we can ensure AI technologies serve humanity positively and responsibly..

Chapter 3: Regulatory Landscape and Emerging Policies

First, the EU AI Act and how it impacts both the European Union and the global AI industry.

Then, the US National Institute of Standards and Technology AI Risk Management Framework.

We'll also examine China's AI governance directives and their focus areas.

Finally, we'll discuss soft law approaches and how industry self-regulation complements formal government regulation.

The EU AI Act is the first comprehensive regulatory framework for AI in the world.

It classifies AI systems based on risk, ensuring that higher-risk applications are subject to stricter controls.

For high-risk AI, requirements include high-quality data, transparency, and strong human oversight.

Overall, the goal is to build public trust in AI while fostering innovation in the sector.

For organizations operating within the EU, the AI Act emphasizes transparency and accountability.

AI systems must be explainable and subject to clear governance processes to comply with EU standards.

Conformity assessments are now required, ensuring that AI systems meet safety and ethical guidelines before deployment.

Key sectors like healthcare, transport, and finance are particularly affected, meaning they must develop specific strategies to ensure compliance.

The EU AI Act is more than just a regional policy — it sets a precedent for AI regulation worldwide.

By harmonizing standards, it makes it easier for firms to comply with laws in multiple jurisdictions.

International companies are already adjusting their strategies to align with the Act's requirements, seeing it as a gateway to access global markets.

The US NIST AI Risk Management Framework was developed to help organizations manage AI risks effectively.

It focuses on principles like transparency, fairness, accountability, and robustness.

The framework offers practical tools for assessing AI risks and implementing mitigation strategies.

It also stresses the importance of continuous monitoring to adapt to evolving risks and organizational needs.

The NIST Framework is being adopted across a range of industries:

In healthcare, it supports AI reliability and patient safety.

In finance, it's used for risk management and fraud detection.

In defense, it helps ensure ethical use of AI and strengthens security.

Beyond industry, the Framework is influencing national AI policies and helping shape standards development in the US.

Chapter 3: Regulatory Landscape and Emerging Policies

In China, AI governance is led by key governmental bodies like the Ministry of Science and Technology and the Cyberspace Administration.

They have established regulations and ethical guidelines to govern AI development and deployment. China's national AI strategies aim to drive innovation while promoting responsible AI use across different sectors. China's AI governance focuses on three main areas:

First, promoting ethical AI use to ensure fairness and prevent misuse.Second, enhancing cybersecurity to protect AI systems and sensitive data from threats. And third, enforcing strict data privacy regulations to protect personal information and build trust with users.

Soft law refers to voluntary guidelines and standards that aren't legally binding but influence industry behavior.

Examples include ethical codes of conduct and industry-developed best practices, which can guide responsible AI development even without formal regulation.

Industry coalitions and standards bodies play a key role in shaping soft law. They develop voluntary standards, promote ethical AI practices, and encourage collaboration among companies, researchers, and regulators.

These efforts help build a consensus on what responsible AI governance should look like across different industries. Soft law has both strengths and limitations. Its main advantage is adaptability — it can respond quickly to changes in AI technology.

However, because it lacks formal enforcement, its effectiveness depends on voluntary compliance.

Soft law works best when it complements formal regulation, filling gaps and promoting industry responsibility in AI governance.

Chapter 3: Regulatory Landscape and Emerging Policies

To wrap up, the global AI regulatory landscape is rapidly evolving. We've seen how the EU, the US, and China are shaping AI governance through laws, frameworks, and strategies. At the same time, soft law and industry self-regulation are playing vital roles in promoting ethical AI use.

Understanding this dynamic environment is crucial for organizations that want to innovate responsibly and stay compliant in a complex global market.

Chapter 4: AI Risk Management

AI risk management involves several key elements. First, we'll look at how to identify and classify AI risks. Then, we'll cover AI Risk Impact Assessments, known as AIRIA, and how they're used. We'll also discuss how to put guardrails and controls into operation and finally, how to handle incident response and harm mitigation.

AI risk management frameworks play a critical role in recognizing potential risks early on — whether ethical, operational, or security-related. These frameworks offer structured processes that help organizations systematically assess the severity and impact of AI-related risks. They also provide guidance on effective mitigation strategies to control and reduce those risks.

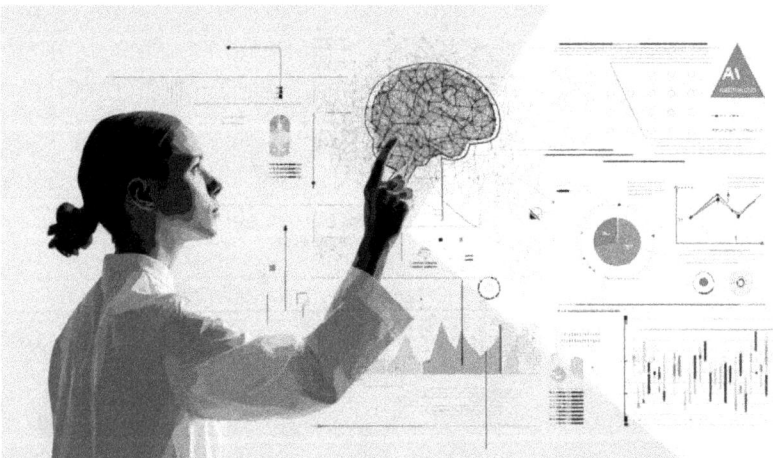

Managing risk in AI is essential for three main reasons. First, it helps prevent harm by identifying and addressing risks before they become real problems. Second, it ensures that AI systems comply with both legal requirements and ethical standards, reducing regulatory risks. And third, good risk management builds public trust by making AI systems reliable, transparent, and fair.

AI risk management involves multiple stakeholders — developers, risk managers, executives, and regulators all need to work together. Clear roles and responsibilities are critical to ensure accountability in this process. Collaboration mechanisms, like structured communication channels, make this teamwork more effective and ensure comprehensive oversight.

There are several methods we can use to recognize AI-related risks. Scenario analysis helps us evaluate potential risks by considering different future outcomes. Audits provide a systematic review of AI systems to uncover weaknesses. And stakeholder interviews gather insights from various perspectives, giving us a broader understanding of potential risks.

Chapter 4: AI Risk Management

Classifying risks helps us manage them effectively. We start by identifying the type of risk, whether ethical, operational, or technical. We then assess the severity — understanding the potential impact allows us to prioritize. Finally, we estimate the likelihood of each risk occurring so that we can allocate mitigation resources appropriately.

Effective risk documentation is vital. Comprehensive risk registers allow teams to track, update, and communicate risks consistently. Digital risk management tools enhance this process by enabling real-time updates and fostering better collaboration. Proper documentation also ensures transparency and prepares the organization for audits and external reviews.

AI Risk Impact Assessments, or AIRIA, have a clear objective — to systematically assess ethical, legal, and operational risks. Stakeholder mapping is an essential part of this process, helping us identify key players and their concerns. Impact scoring lets us prioritize risks based on potential effects. And scenario testing allows us to anticipate and prepare for different outcomes.

When assessing the potential impact on stakeholders, we first identify all individuals, communities, or organizations that may be affected. It's important to consider both direct and indirect effects. Once risks are identified, we tailor mitigation strategies to address them while also enhancing positive outcomes for all stakeholders involved.

Integrating AIRIA into AI development cycles helps manage risks proactively. By embedding these assessments early in a project, we can address risks before they escalate. Incorporating AIRIA iteratively supports continuous improvement, and ensuring ethical alignment throughout development promotes responsible innovation.

AI guardrails are designed to prevent harm and ensure ethical operation. They need to strike a balance between maintaining safety and allowing for effective system performance. Well-designed guardrails help AI systems function within acceptable boundaries while still delivering value.

We implement both technical and procedural controls to manage AI risks. Technical controls include algorithms and automated monitoring systems. Procedural controls consist of policies, guidelines, and human oversight. Using both together creates a stronger, more resilient risk management framework.

Guardrails aren't a one-time setup. Continuous monitoring helps us detect deviations and emerging risks as AI systems evolve. We also need to update guardrails regularly to adapt to changes in AI capabilities and the operating environment, maintaining ongoing effectiveness.

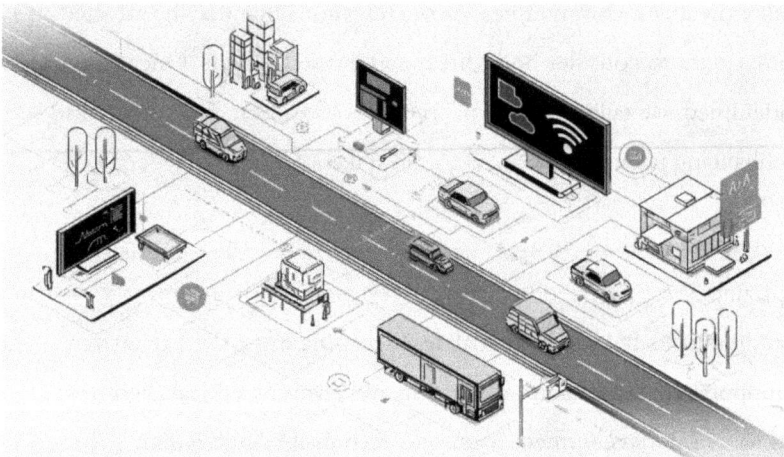

Effective incident response protocols start with clear detection steps, using monitoring and alert systems to spot problems early. Defined escalation procedures ensure that the right people are informed quickly. Having a clear resolution and remediation plan allows teams to contain and resolve incidents efficiently.

Communication is key during AI-related incidents. Transparent stakeholder communication helps maintain trust and provides clarity in challenging situations. Accurate reporting supports accountability and ensures organizations meet both internal and external regulatory requirements.

After an incident, it's important to conduct a root cause analysis to understand what went wrong and why. These insights drive improvements in both systems and processes, enhancing overall safety and efficiency. Continuous learning from incidents also strengthens future risk management practices.

To wrap up, AI risk management is a continuous, proactive process that spans risk identification, assessment, control implementation, and incident response. By following these strategies, organizations can deploy AI systems responsibly, ensuring safety, compliance, and trust in their operations.

Chapter 5: Trustworthy AI Design

Building trustworthy AI isn't without challenges. Managing bias is crucial because biased AI systems can lead to unfair outcomes. Ensuring interpretability helps users understand how AI makes decisions. We must also protect user data and comply with privacy regulations. Finally, robustness is key—AI needs to withstand adversarial attacks and perform reliably.

Let's start with the difference between explainability and interpretability. Interpretability is about understanding how the AI works inside—how it processes data and makes decisions. Explainability, on the other hand, focuses on explaining why the AI made a specific decision or prediction in a way that humans can understand.

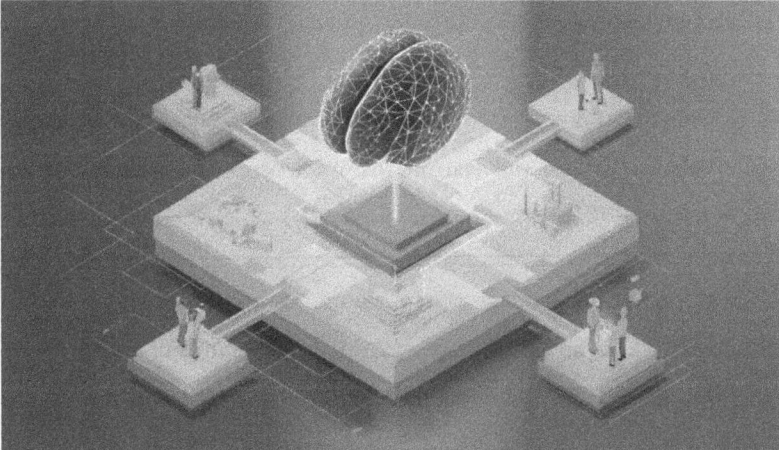

There are several methods to make AI models more transparent. We can simplify models to make them easier to interpret without losing much performance. Feature importance analysis tells us which input variables most influenced the AI's decisions. Visualization tools can help show how the AI behaves. And local explanation methods, like LIME and SHAP, explain individual predictions in a meaningful way.

So why does this matter? Greater transparency builds trust with users and helps them feel confident in the AI's decisions. Explainable AI also helps with regulatory compliance by providing clear audit trails. In sectors like healthcare, finance, and law, explainable AI supports better decision-making and risk management.

AI systems often use sensitive personal data, making privacy a key concern. We need proper data handling practices and always obtain user consent. It's also critical to apply technical safeguards to prevent unauthorized access or misuse of data in AI systems.

Two common techniques for privacy-preserving machine learning are federated learning and differential privacy. Federated learning allows AI models to train on local devices without transferring sensitive data.

Chapter 5: Trustworthy AI Design

Differential privacy works by adding statistical noise to data, protecting individual privacy while keeping the data useful.

The real challenge is balancing data utility with privacy protection. We want AI systems that are both effective and respectful of users' privacy, finding the right trade-off between useful insights and safeguarding personal information.

AI systems face serious security threats. Data poisoning attacks involve corrupting training data to damage AI performance. Adversarial attacks trick AI models with manipulated inputs. Model theft happens when someone copies an AI model without permission. We need strong mitigation strategies to protect AI integrity and reduce these risks.

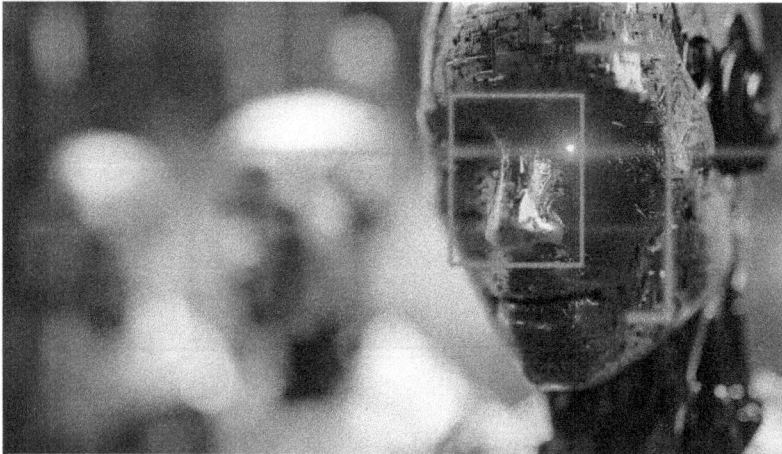

Some best practices for secure AI development include: following secure coding standards to reduce vulnerabilities, rigorous testing to uncover weaknesses before launch, using access controls and encryption to protect data and models, and integrating security early in the development lifecycle to ensure resilience from the start.

Security doesn't stop at deployment. We need continuous threat detection to catch new risks early. Regular risk assessments help us manage security proactively and ensure we stay compliant with regulations.

Inclusive design starts with diversity in AI teams and data. Diverse teams bring varied perspectives, reducing bias and making AI systems fairer and more accurate. And using diverse datasets helps AI better represent real-world populations and scenarios.

We also need to engage stakeholders and affected communities. Inclusive collaboration brings in different voices and ensures AI solutions meet real-world needs. Transparency in the design process builds trust, and involving communities helps ensure the relevance and acceptance of AI systems.

Bias in AI can't be solved with a one-time fix. Bias audits systematically check for discriminatory patterns. Fairness-aware algorithms help correct biases during development. And ongoing evaluation ensures the AI remains fair as new data comes in over time.

To wrap up, trustworthy AI requires a holistic approach. It's about ethics, transparency, fairness, privacy, security, and inclusivity. By following these principles and best practices, we can build AI systems that earn trust and deliver positive outcomes for everyone.

Chapter 6: Bias, Discrimination, and Fairness

Let's begin by defining three critical terms. Bias in AI refers to systematic errors or prejudices that skew decision-making processes. Discrimination happens when AI systems treat individuals or groups unfairly based on these biases. And fairness is the principle of ensuring equitable treatment and outcomes for everyone affected by AI applications. These are foundational concepts in developing responsible AI systems.

Why do these concepts matter? The use of AI can have serious ethical and societal implications. If bias and discrimination go unchecked, AI can reinforce harmful stereotypes, make unfair decisions, or perpetuate inequalities. Understanding and addressing these risks is essential for responsible AI deployment.

AI bias often stems from the data itself. Biases can be introduced through sampling errors, biased labeling, or poor representation of certain groups within datasets. If the data is flawed, the AI model built on it will inherit those flaws—sometimes amplifying them in decision-making.

Bias can also come from how algorithms and models are designed. The choices developers make—whether consciously or not—can embed bias into the system. Optimization criteria may favor certain groups over others, and a lack of transparency can make it hard to spot or fix these issues later. Fairness needs to be considered right from the design stage.

Bias doesn't just exist in data or algorithms. It can also arise from broader societal and systemic factors or how AI is deployed. Deployment contexts, institutional practices, and existing inequalities can all shape how bias appears and affects people in real-world applications.

Detecting bias in AI requires a range of techniques. Statistical analysis helps us spot imbalances or unusual patterns in data. Fairness testing allows us to check whether AI models treat different groups equitably. And visualization tools make disparities easier to see and understand. These methods are essential first steps in managing AI bias.

Beyond detection, we need robust auditing and evaluation. Fairness toolkits can assist developers in identifying and mitigating bias. Independent third-party audits can validate whether AI systems perform fairly. And systematic evaluation frameworks help ensure fairness is an ongoing priority, not just a one-time check.

However, measuring bias isn't always straightforward. Different metrics may suggest different conclusions, and interpreting them requires care. Bias detection is a complex process, and drawing the right conclusions depends on understanding the context, limitations, and nuances of each method used.

Several fairness metrics are commonly used to evaluate AI models, like demographic parity or equalized odds. These metrics help us assess whether a model's decisions are equitable across different groups. However, each metric captures fairness in a different way, and no single metric tells the full story.

Achieving fairness often involves trade-offs. Sometimes, improving fairness may slightly reduce a model's accuracy or utility. Balancing these factors is a key challenge in AI development. Ethical considerations must guide these decisions to ensure AI serves the best interests of all stakeholders.

Another challenge is that fairness metrics can conflict. You might be able to satisfy one metric but fail another. This makes defining fairness a complex, sometimes subjective task. There's rarely a one-size-fits-all solution when it comes to fairness in AI.

These conflicting definitions also make practical implementation difficult. Developers have to navigate competing fairness goals and balance them with technical constraints. It's a nuanced process requiring careful thought and a strong ethical framework.

So, how can we reduce bias and promote fairness? Data augmentation helps by making training data more diverse. Algorithmic adjustments can help correct biased decision-making. And fairness-aware learning integrates fairness into the training process itself, promoting equitable outcomes from the start.

It's also vital to give individuals affected by AI decisions options for recourse. They need ways to challenge decisions, seek corrections, or receive compensation if harmed. This promotes transparency, trust, and accountability in AI systems.

Policy, regulation, and governance frameworks play a huge role in ensuring fairness. Clear legal standards guide ethical AI development. Governance models provide oversight and accountability. And strong policies help protect society, promoting fairness and building public trust in AI technologies.

In conclusion, understanding bias, discrimination, and fairness in AI is critical for building responsible systems. We've seen how bias can emerge from data, algorithms, or deployment, how we can detect and address it, and why fairness is both essential and challenging. By applying ethical principles, using sound evaluation methods, and establishing robust governance, we can work toward more equitable AI systems that benefit everyone.

Chapter 7: Corporate AI Governance Structures

AI governance in a corporate context refers to the systems and policies organizations put in place to guide the ethical development and use of AI. The goal is to ensure AI is created and deployed responsibly, reflecting societal values and ethical standards. At the same time, AI governance ensures that AI initiatives are aligned with the business's overall objectives, supporting both innovation and accountability.

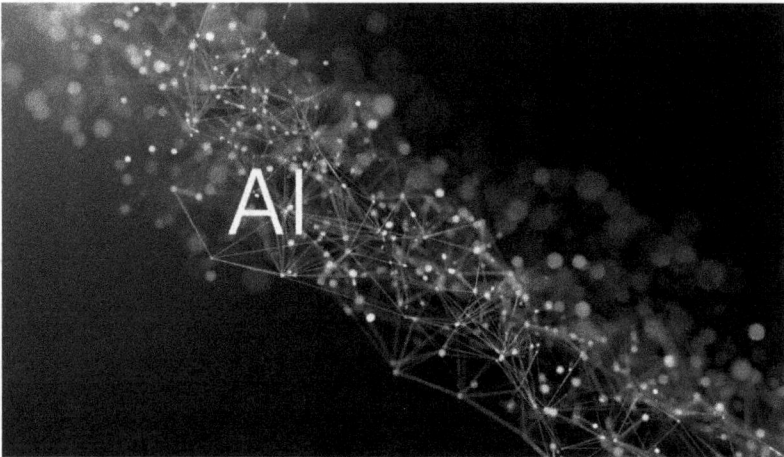

Structured AI governance is crucial for several reasons. First, it helps mitigate risks related to AI deployment and operation, safeguarding both the company and its users. It also ensures that organizations comply with regulations, helping them avoid legal penalties and reputational damage. Beyond compliance, structured governance builds transparency, which fosters trust among customers and stakeholders. Finally, it promotes responsible AI management throughout the technology's lifecycle, embedding ethics and accountability at every stage.

Chapter 7: Corporate AI Governance Structures

Corporate AI governance frameworks typically include several key components. Risk management models are used to identify and reduce potential threats posed by AI systems. Ethical guidelines provide foundational principles for responsible AI development. Compliance checklists help organizations ensure their AI adheres to relevant laws and standards. And finally, oversight committees—often cross-functional—are established to monitor AI governance and ensure accountability across the organization.

AI ethics boards play a critical role in corporate governance. They provide independent oversight to help maintain ethical standards during AI development and deployment. These boards evaluate potential ethical risks associated with AI projects, ensuring that innovation doesn't come at the expense of ethical responsibility. They also support decision-making within the organization, guiding leaders through complex ethical considerations in AI initiatives.

The composition of AI ethics boards is vital. They should include a diverse mix of experts—technical specialists, ethicists, legal advisors, and business leaders. This diversity ensures a well-rounded perspective when evaluating AI projects. The decision-making processes of these boards should be clear and transparent, often informed by real case studies or precedents, to ensure that their guidance is practical and grounded in real-world applications.

The Chief AI Ethics Officer has several key responsibilities. They oversee AI projects to ensure compliance with ethical standards, helping to prevent bias and harm. They also lead the development and enforcement of governance policies related to AI. Importantly, this officer acts as a liaison between technical teams and business executives, ensuring that ethical considerations are integrated into business strategies and objectives.

AI governance officers don't work in isolation. They collaborate with various teams across the organization, ensuring that governance practices are built into both AI development and operational processes. This collaboration draws on a wide range of expertise—from data scientists and engineers to legal experts and business units—to create a comprehensive approach to AI governance.

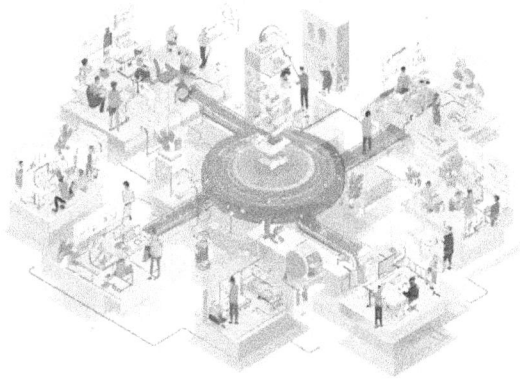

Accountability is a key part of AI governance. Governance officers need clear reporting structures to ensure transparency and traceability in AI projects. These structures help organizations maintain oversight and take corrective action when needed, reinforcing ethical standards and accountability across all levels.

Integrating governance into MLOps pipelines is critical for ethical AI deployment. Embedding ethical checks ensures AI models align with moral and societal standards. Compliance validations throughout the MLOps process help guarantee adherence to regulations. Finally, implementing audit trails creates a transparent record of AI model development and deployment, enhancing accountability.

AI governance must be present at every stage of the AI lifecycle. During data collection, risk assessments help ensure data privacy and quality. In model development, mitigation strategies are applied to reduce bias and errors. At deployment, monitoring ensures ongoing compliance with ethical standards. And even at the decommissioning phase, risk controls are necessary to manage data security and other concerns effectively.

Continuous monitoring and feedback mechanisms are essential for maintaining AI integrity. Real-time monitoring helps organizations quickly detect and address performance issues or ethical deviations. Feedback loops enable timely adjustments and corrections, ensuring that AI systems remain aligned with organizational values and ethical standards throughout their operation.

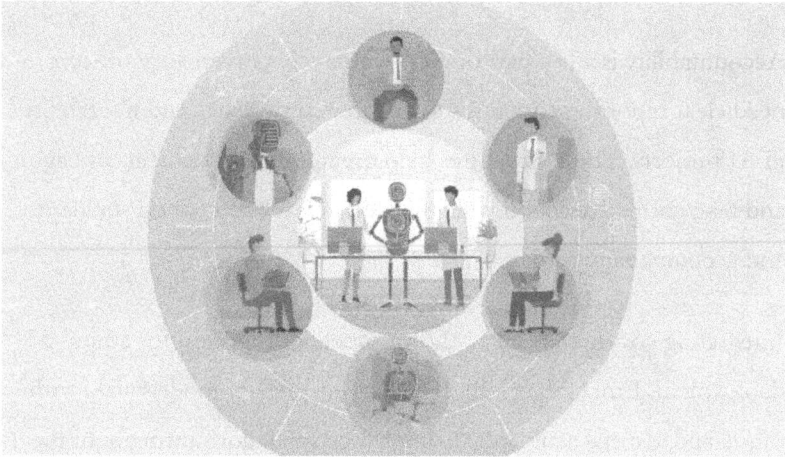

Establishing clear AI policies and guidelines is the foundation of effective governance. These policies should define acceptable AI practices in alignment with the company's goals. They must also embed ethical principles, ensuring fairness and responsibility in AI use. Finally, clear compliance requirements help organizations meet the necessary legal and regulatory standards related to AI.

AI policies aren't static—they need regular review and updates. Engaging stakeholders in this process brings in diverse viewpoints and fosters collaboration. Regulatory scanning helps organizations keep up with changes in laws and compliance standards. And conducting impact assessments ensures that policies are effective and highlight any areas that might need improvement.

It's essential that AI policies stay in harmony with both regulations and evolving ethical standards. Ongoing policy harmonization reduces legal risks and helps the organization uphold its corporate responsibility. By aligning policies with both legal requirements and ethical expectations, companies can maintain trust and act responsibly in the AI space.

To conclude, effective corporate AI governance structures combine clear frameworks, defined roles, and robust processes. By embedding ethics, accountability, and compliance into every stage of AI development and deployment, organizations can not only minimize risks but also build trust with stakeholders and drive responsible innovation.

Chapter 8: AI Auditing and Assurance

AI auditing is about ensuring that AI systems function transparently, ethically, and reliably. Given how AI impacts decisions that affect people and organizations globally, auditing is essential. It builds trust by holding AI systems to high standards and verifying their behavior in critical applications.

The main objectives of AI assurance are threefold. First, ensuring ethical compliance—making sure AI aligns with agreed-upon ethical standards. Second, guaranteeing reliability and accountability—so that systems work as intended and creators take responsibility for outcomes. And third, risk mitigation—identifying and reducing potential risks when AI is deployed.

Auditing AI systems comes with unique challenges. AI algorithms are often highly complex, making their decisions hard to audit. Poor or biased data can undermine audit results. The lack of standardized metrics complicates consistent evaluations, and rapidly changing regulations make it hard to stay compliant. These factors all make AI auditing a demanding task.

AI audit approaches typically fall into three categories. Technical assessments check the algorithms and data for accuracy and reliability. Compliance checks ensure systems meet industry regulations. And ethical evaluations examine impacts like fairness, bias, and societal influence. Each plays a key role in comprehensive AI auditing.

Risk-based audits focus on the AI components that carry the highest risk or potential impact. This allows organizations to prioritize resources on areas that need the most oversight, ensuring critical aspects of AI systems get the attention they require for effective assurance.

There are several key techniques used to assess AI algorithms and data. Algorithm performance testing measures how accurately and efficiently the AI performs. Bias detection looks for and addresses unfair outcomes. Training data validation checks the quality and relevance of the data used to train AI models. And model behavior monitoring ensures the AI behaves as expected over time.

Third-party audits provide an independent check on AI systems, which helps build credibility and transparency. They enhance trust among stakeholders by confirming compliance with ethical standards and regulations. These audits ensure AI systems meet both legal and industry requirements, serving as a critical part of governance.

Internal audits focus on continuous monitoring of AI systems within the organization. They help manage risks, ensure ongoing compliance, and support the improvement of AI processes over time. These audits offer a valuable internal check on how AI systems are operating day to day.

Chapter 8: AI Auditing and Assurance

Third-party and internal audits serve different but complementary purposes. Third-party audits offer independent validation and boost external credibility. Internal audits provide deeper insights into operations and help fine-tune internal controls. Together, they create a strong foundation for AI assurance and effective risk management.

Audit trails are crucial for transparency in AI systems. They create a record of decisions, making it easier to trace how AI systems reach conclusions. These trails support accountability by documenting system changes and help investigations by providing clear records when issues arise.

Chapter 8: AI Auditing and Assurance

Logging standards specify what data AI systems need to capture for auditing. Logs must be stored securely to protect their integrity, and it's vital that they are tamper-proof. This ensures audit records remain reliable and useful for compliance and investigation purposes.

Maintaining audit logs effectively means conducting regular reviews to spot anomalies, using secure storage with encryption, and setting strict access controls. It's also important to integrate audit logs into incident response processes to quickly address any security issues that arise.

AI certification schemes formally recognize systems that meet ethical, technical, and governance standards. They evaluate AI systems on multiple criteria and help build trust among users and stakeholders by demonstrating that the system complies with recognized benchmarks.

ISO/IEC 42001 outlines comprehensive management system requirements for trustworthy AI development and deployment. It emphasizes strong risk management, promotes transparency, and ensures accountability in AI systems. These principles help organizations develop AI responsibly.

Certification offers clear benefits, such as increased trust and better market acceptance. However, it also comes with challenges. AI standards are evolving quickly, making it hard for certifications to stay relevant. Meeting certification requirements can be costly, and rapid AI innovation makes it tough for certification schemes to keep pace.

To conclude, AI auditing and assurance are essential for trustworthy AI deployment. They provide transparency, uphold ethical standards, and ensure reliability. By combining rigorous audits, effective logging, and adherence to certification schemes, organizations can manage AI risks effectively and foster confidence among stakeholders.

Chapter 9: Responsible Data Governance

Responsible data governance is about handling data in a way that aligns with ethical standards. This ensures individuals' privacy is respected and protected. It also means organizations remain accountable for how they manage data, ensuring transparency and building trust with stakeholders. A strong governance approach safeguards against unauthorized access and misuse, maintaining integrity at every step.

Key principles of governance include transparency—so people know how their data is used—accountability, ensuring organizations take ownership of data practices, maintaining high data quality, securing data properly, and staying compliant with laws and regulations. Frameworks like GDPR and ISO standards help organizations put these principles into action, providing structured guidance for effective governance programs.

Chapter 9: Responsible Data Governance

Implementing governance comes with real challenges. First, there are complex regulatory environments—laws differ by region and change over time. Data silos within organizations make it hard to apply governance consistently. Limited budgets and staff can constrain governance efforts. And finally, finding the right balance between making data accessible for business use while keeping it secure is always a tough act.

Data provenance is all about understanding where data comes from and its history. This information is critical for ensuring data can be trusted and validated. Knowing the provenance also helps organizations comply with regulations and make informed decisions because they have a clear view of the data's background and context.

Data lineage is closely related—it tracks the path data takes through an organization, including any changes along the way. This helps organizations see how data moves, where it originates, what systems it interacts with, and what transformations it undergoes. Knowing this allows organizations to manage risks, dependencies, and maintain integrity in their data systems.

Organizations can use automated lineage tools to map data flows across systems, making tracking easier and more accurate. Managing metadata effectively also plays a big role, as it gives context to the data, supporting traceability. And, of course, having clear data policies ensures consistent practices and reinforces governance goals throughout the organization.

Chapter 9: Responsible Data Governance

Consent management is built on a foundation of legal requirements—especially data protection laws like GDPR—that require organizations to obtain clear and informed consent from individuals before using their data. Consent must be specific, freely given, and documented, ensuring that organizations remain compliant and trustworthy in their data practices.

To manage user consent effectively, organizations should use clear, transparent consent forms that people can easily understand. Offering granular consent choices lets users decide exactly how their data is used. Consent management systems help track, update, and maintain these consents over time, ensuring that everything stays compliant and up to date.

Empowering individuals means giving them access to their personal data whenever they ask for it. This transparency builds trust. It's also important to allow users to correct inaccuracies in their data. These mechanisms not only improve data quality but also give users more control over their personal information, reinforcing good governance practices.

Synthetic data is data that's artificially generated to resemble real data. It's particularly useful in AI and data testing environments where using real data may pose privacy risks. By using synthetic data, organizations can train models and run analyses without exposing sensitive information, providing both flexibility and a layer of privacy protection.

Synthetic data offers strong privacy benefits since it replaces real sensitive data in many use cases. It also increases data availability, making it easier to train AI models or run tests. However, synthetic data has its limitations—it may not fully capture real-world complexities, which can affect accuracy and reliability. Plus, if it's not representative of real populations, it could impact fairness and generalizability of models.

Chapter 9: Responsible Data Governance

Data minimisation means collecting only the data you really need. This reduces privacy risks and helps with compliance. It's also essential to have policies in place for deleting data when it's no longer needed—this helps prevent unnecessary exposure. Overall, minimising data collection and retention helps lower the risk of breaches and strengthens data security.

Data sovereignty means that data is subject to the laws of the country where it's stored. This affects how organizations manage data, especially when dealing with global operations. Understanding local regulations is crucial, as failing to comply can have serious legal consequences. It's a key factor when setting up governance strategies.

Cross-border data transfers come with their own challenges. Different countries have varying laws on how data can move across borders. Organizations need to understand these rules to manage risk and ensure compliance. Navigating these complex legal landscapes is a major part of responsible data governance.

Organizations must have strong compliance programs in place to manage international data flows securely and legally. Frameworks like GDPR and CCPA offer guidance on maintaining data privacy across borders. Managing data in this way not only keeps organizations compliant but also protects user data and upholds trust.

To wrap up, responsible data governance is a multifaceted approach that demands attention to ethical standards, legal requirements, and operational best practices. By focusing on the principles we've discussed—like data provenance, consent management, synthetic data use, and respecting data sovereignty—organizations can build stronger, more trustworthy data management programs that protect both individuals and themselves.

Chapter 10: Ethics in Generative AI

Ethics in generative AI is about more than just rules — it's about understanding the responsibilities that come with creating and using AI systems that can generate content. We need to ask how these systems impact individuals, communities, and society at large, and how we can ensure they are used in ways that align with our values and respect for human dignity.

Generative AI has advanced rapidly, moving from simple algorithms to models that can produce highly realistic content. Understanding this historical evolution is important because it gives context to the capabilities—and the ethical questions—we face today. As these models become more powerful, new ethical challenges emerge, like content manipulation, misinformation, and identity misuse, all of which require thoughtful consideration.

Several key stakeholders play a role in shaping ethical standards for generative AI. Developers have a responsibility to design systems with ethical principles built-in. Users also have a duty to apply AI responsibly. Regulators create the legal frameworks that guide and control AI usage, while society as a whole influences norms and holds all these players accountable for the broader impact of AI.

Deepfakes are a prime example of generative AI's power—and its risks. These are AI-generated images or videos that can convincingly imitate real people. While they can be used for creative purposes, they also present serious threats when used maliciously, such as spreading false information or damaging reputations. That's why awareness and vigilance are critical.

Voice cloning poses a similar threat. AI can now replicate voices with remarkable accuracy, making identity theft and fraud easier to carry out. Misuse of this technology can lead to serious societal harm, which is why it's crucial to raise awareness and implement protective measures to prevent abuse.

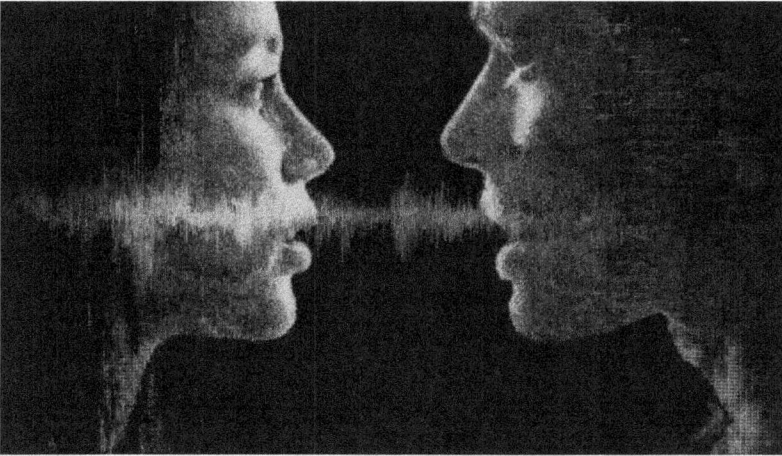

To combat manipulation and verify content authenticity, forensic analysis can be used to examine digital traces and metadata. Alongside this, AI-powered detection tools are emerging that can help identify synthetic media, offering a technical safeguard against malicious use. Both forensic techniques and AI detection are important tools in ensuring content integrity.

AI-generated content raises complex questions about intellectual property. When a piece of content is created by a combination of algorithms, data inputs, and minimal human input, who really owns it? This gray area challenges traditional legal concepts and calls for updated frameworks that reflect the realities of AI-generated works.

Attribution is tricky with AI-generated outputs. Is it the user, the AI developer, or the data provider who owns the rights? And who is responsible if something goes wrong? These questions highlight the difficulty in managing rights and responsibilities, both legally and ethically, when dealing with AI-generated content.

Laws are trying to catch up with AI's rapid growth. Existing frameworks are evolving, with new regulations and policies aiming to balance the need for innovation with the protection of individual and societal rights. It's a delicate balance—one that requires constant adaptation and thoughtful policy-making.

We need strong ethical guidelines and governance for large language models and foundation models. These guidelines ensure that AI systems operate with transparency, fairness, and respect for human rights. Governance structures play a key role in providing oversight and accountability to prevent misuse.

On a technical level, there are several ways to implement safeguards. Content filtering helps block harmful outputs. Bias mitigation techniques work to reduce unfairness in AI responses. Usage monitoring allows developers to track how AI is being used, which can help prevent abuse and maintain ethical standards in practice.

Chapter 10: Ethics in Generative AI

Balancing innovation with risk is essential. While we want to encourage creativity and advancement in AI, we also have to manage the risks that come with it. Effective governance ensures that AI continues to grow in a responsible way, promoting progress while safeguarding against harm.

Content moderation is critical when dealing with generative AI. Automated moderation tools can efficiently scan for harmful or inappropriate content. But human reviewers are still essential, especially for making nuanced judgments that automated systems might miss. The best approach often combines both human oversight and AI tools for effective moderation.

Generative AI can sometimes introduce bias, misinformation, or harmful content. It's important to identify and reduce bias, implement processes for verifying information, and develop safeguards to minimize potential harm. These efforts help ensure that AI serves society in a fair and responsible manner.

Chapter 10: Ethics in Generative AI

Generative AI is reshaping society, bringing both opportunities and challenges. Ethical stewardship is necessary to guide its development and application. At the same time, policies must remain flexible and adaptive to keep pace with AI's rapid advancement and the new issues it presents.

To conclude, generative AI presents incredible possibilities, but it also brings serious ethical considerations. By understanding the risks, embracing responsible practices, and promoting effective governance, we can harness the power of generative AI for good—while minimizing its potential for harm.

Chapter 11: Human Oversight and AI Control

Let's begin by defining Human-in-the-Loop, or HITL. This approach integrates human judgment directly into the AI decision-making process, making outcomes more accurate and reliable. For example, in medical diagnosis, HITL combines AI's data analysis with expert human evaluation to provide better patient outcomes. In autonomous vehicles, HITL allows human oversight to guide decisions in real time, enhancing safety. Similarly, in critical military operations, HITL ensures human judgment directs AI actions during sensitive missions.

Now, let's explain Human-on-the-Loop, or HOTL. In this model, humans oversee AI systems but only intervene when needed. Unlike HITL, HOTL reduces the need for continuous human involvement, making it more scalable for routine tasks. However, HOTL relies on strong monitoring systems that can quickly detect issues and allow human operators to step in before problems escalate.

Chapter 11: Human Oversight and AI Control

When comparing HITL and HOTL, the task's complexity often determines the right approach. HITL is better for complex tasks where human insight is essential, while HOTL works well for routine operations with lower risk. HOTL can also offer faster system responses since humans are not constantly involved, but only step in as supervisors when necessary.

Override capabilities are critical for human control over AI. Emergency stop functions allow operators to immediately halt AI operations during unexpected or harmful behavior. Manual control interfaces give humans direct control over the AI system when necessary. Additionally, override commands built into AI interfaces let operators modify or overrule AI decisions quickly when required.

Failsafe systems are designed to ensure safety when AI fails. These include fallback operations that activate if the AI malfunctions. Building redundancy into AI design ensures that backup components take over if primary systems fail. Error detection mechanisms catch problems early, and safe state transitions help move systems into secure modes during malfunctions to prevent harm.

Chapter 11: Human Oversight and AI Control

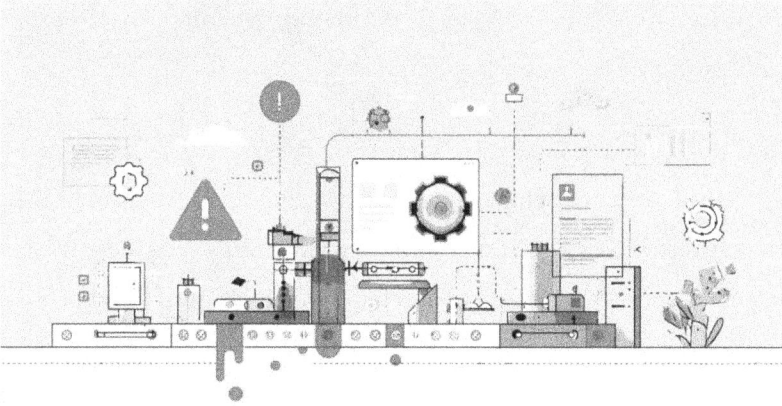

Let's look at practical examples. In autonomous vehicles, override and failsafe systems protect passengers during unexpected situations. In industrial robotics, failsafes prevent accidents and protect workers. Healthcare AI systems use override functions to avoid errors that could compromise patient safety. These examples show how critical human oversight and robust control mechanisms are in safety-critical environments.

Accountability in AI means clearly defining who is responsible for decisions and outcomes. This includes AI developers, system operators, and the organizations that deploy these systems. Traceability mechanisms, like audit trails and logs, make AI decisions transparent and help clarify responsibility when things go wrong.

Legal regulations ensure AI systems are compliant and that accountability is enforceable. Ethical guidelines support responsible AI design, promoting fairness, transparency, and reducing bias. Together, legal and ethical frameworks guide the development of AI systems that are trustworthy and safe for society.

Real-world case studies highlight the challenges of AI accountability. These often involve complex situations where it's hard to pinpoint responsibility due to AI's autonomous nature. Addressing these issues requires a mix of legal regulations, ethical standards, and transparent system design. Learning from these cases helps shape better governance and accountability practices for AI in the future.

Monitoring AI systems in real time involves several methods. Automated alerts notify operators instantly when something goes wrong. Performance dashboards provide a real-time view of system metrics, helping teams monitor and assess AI behavior. Anomaly detection algorithms are critical for spotting unusual activity that could indicate a problem.

Escalation protocols outline how and when issues should be reported to higher-level authorities or operators. Having clear protocols ensures quick action when risks emerge, helping prevent harm and protect the integrity of AI systems.

Human operators play a vital role in ongoing AI oversight. They interpret complex monitoring data, assess risks, and make expert decisions about when to intervene. This human judgment is essential for managing critical issues that automated systems may not fully understand.

Chapter 11: Human Oversight and AI Control

To conclude, maintaining human oversight and control is essential for the ethical and safe use of AI. Whether through active involvement, emergency overrides, or continuous monitoring, human judgment remains central to responsible AI deployment. By applying the principles we've discussed—oversight models, failsafes, accountability, and monitoring—we can ensure AI systems are trustworthy and aligned with human values.

Chapter 12: Ethical Deployment in Specific Sectors

When we talk about ethical deployment in a technological context, we're talking about three main ideas. First, respect for human rights—this means ensuring technology upholds the dignity and rights of every person. Second, promoting fairness—making sure technology doesn't introduce bias and that everyone has equal access. And third, transparency and accountability—organizations must be open about how their technology works and take responsibility for its impact.

Across all industries, there are some core ethical principles we need to uphold. Respecting privacy is crucial—this means protecting people's data and upholding their right to privacy. Fairness and equity means avoiding bias and promoting equal treatment. Transparency and accountability involve clear communication and taking responsibility for how technology is used. Finally, preventing harm means proactively identifying and managing risks that could arise from technology deployment.

Chapter 12: Ethical Deployment in Specific Sectors

When it comes to specific sectors, we see both challenges and opportunities. Ethical challenges include risks like bias, data privacy issues, and accountability gaps—all of which can erode trust. On the other hand, responsible technology use can drive innovation, improve efficiency, and promote fairness across sectors if handled properly.

Patient privacy is a top priority in healthcare. Protecting patient information builds trust and maintains confidentiality. Compliance with regulations like HIPAA is essential for securing patient data. Healthcare organizations must implement strong security measures, such as encryption and strict access controls, to safeguard sensitive medical information from breaches.

Diagnostic algorithms must be carefully designed to minimize bias that could affect diagnostic accuracy and patient outcomes. Ensuring fairness in these tools helps avoid misdiagnosis and ensures that all patients—regardless of background—receive equitable care.

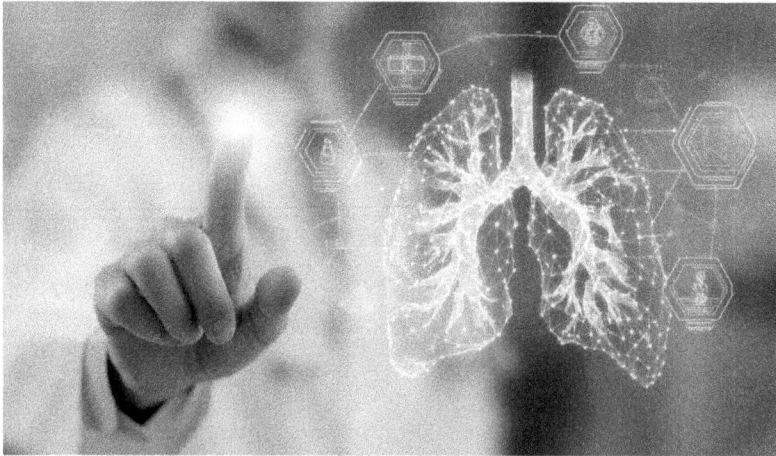

In clinical decision support, accountability is critical. Healthcare professionals must take responsibility for the outcomes influenced by these systems. Ethical guidelines help ensure these tools are used appropriately and always in the best interest of patients.

In financial services, transparency in trading algorithms helps build trust with regulators and stakeholders. It's important that these algorithms are explainable—so that decision-making processes are clear and unintended consequences can be avoided.

Managing systemic risk is essential in automated trading. Tools must be in place to detect early signs of market instability. Additionally, strong monitoring systems are needed to prevent manipulative practices that could undermine market fairness.

Automated lending and credit scoring bring their own ethical considerations. Systems must be designed to avoid discrimination and ensure fair access to financial services. Transparency is also vital—

stakeholders need to understand how these algorithms make decisions to maintain trust.

In the public sector, balancing public safety with individual privacy is a constant challenge. Algorithms used by governments must be transparent to foster trust and enable public scrutiny. Oversight mechanisms are crucial to ensure fairness and accountability in public services.

Surveillance systems must be carefully managed to prevent misuse and overreach. Effective safeguards protect civil liberties and prevent unjust intrusion. Importantly, these systems must avoid disproportionately targeting vulnerable groups to ensure justice and fairness.

In education, protecting student data is critical. Schools must comply with privacy laws and implement safeguards to prevent unauthorized access and data breaches. This helps maintain trust within the education system.

Bias in educational analytics is a real risk. Algorithms can inherit biases from data, leading to unfair treatment of students. Thoughtful design of algorithms can help ensure fair analysis and equal opportunities for all students.

Ethical use of technology in education supports fairness and transparency. It also enables personalized learning, which can better meet individual student needs. By making educational technology accessible, we can help bridge opportunity gaps and promote equality.

In conclusion, navigating ethical challenges in technology deployment requires a deep understanding of both the risks and opportunities in each sector. By adhering to ethical principles—like fairness, transparency, and accountability—we can ensure that technology serves the greater good and fosters trust across all areas of society.

Chapter 13: Future Trends in AI Governance

Autonomous AI systems are increasingly being adopted in critical sectors. In healthcare, they're enhancing diagnostics, treatment planning, and patient monitoring, which leads to better outcomes and more efficient care. In finance, AI is automating risk assessments, fraud detection, and trading activities—reshaping the way financial services operate. And in transportation, we see autonomous vehicles and advanced traffic management systems improving safety and efficiency on our roads.

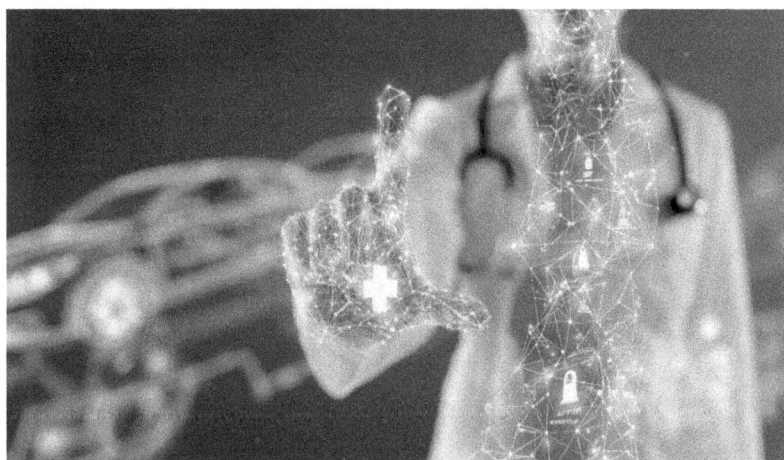

However, with these advancements come governance challenges—questions about who holds decision-making authority, how reliable these systems are, and what societal impacts we need to consider. One of the biggest concerns with autonomous systems is accountability and transparency. These systems often use highly complex algorithms that are difficult—even for experts—to interpret and audit. The decision-making processes are often opaque, making it hard for users and regulators to understand how outcomes are reached. This lack of transparency poses a serious trust issue. For autonomous systems to be accepted and properly governed, we need clear lines of accountability and a foundation of trust between technology providers, regulators, and society.

To address these governance challenges, regulatory frameworks are being developed worldwide. Governments are creating oversight policies to ensure AI is developed and used safely and ethically. Risk management practices are critical to identify and minimize hazards associated with AI systems. And compliance measures make sure that AI technologies meet legal and safety standards, protecting the public's interest while fostering responsible innovation.

AI is a global phenomenon, and cross-border collaboration is key to managing its risks and maximizing its benefits. International cooperation helps promote shared safety and ethical standards, enabling countries to work together on common AI challenges. These agreements aim to strike a balance between encouraging innovation and maintaining effective risk management—an ongoing balancing act for policymakers around the world.

Harmonizing global AI standards is challenging because of jurisdictional and cultural differences. Every country has its own legal system, making unified governance difficult. Cultural values also influence governance approaches, and respecting these differences is vital for developing inclusive policies. Successful governance frameworks manage to balance respect for diversity with the pursuit of shared goals, fostering global cooperation while honoring local values.

As we look toward the future, defining ethical frameworks for advanced AI becomes increasingly important. Ethical principles help ensure that the development of AGI—or artificial general intelligence—aligns with human values and promotes societal well-being. Guidelines for responsible AI development emphasize transparency, safety, and accountability. And strong governance frameworks are essential to oversee and regulate the deployment of superintelligent AI, ensuring it serves humanity's best interests.

Mitigating the risks associated with advanced AI is crucial. Without effective risk control strategies, we risk unintended consequences that could harm individuals or society. Risk mitigation involves proactive planning, continuous monitoring, and the ability to adapt AI behavior when needed. Perhaps most importantly, we must ensure that AI systems are designed to align with human values, promoting ethical use and maintaining public trust.

AI systems don't just affect individual sectors—they have broad societal implications. From reshaping social structures to influencing ethical norms and economic systems, AI's impact is far-reaching. This makes human-AI value alignment even more critical. We must ensure that AI operates in ways that respect human values and contribute positively to society, fostering trust and coexistence between humans and intelligent systems.

As AI technology evolves rapidly, regulations must be adaptable. Rigid rules can stifle innovation, so regulatory models need to be flexible enough to accommodate new advancements. At the same time, these regulations must protect societal values and interests—ensuring that AI development doesn't outpace our ability to manage its risks effectively.

Creating effective AI standards requires collaboration among a wide range of stakeholders. Industry experts, governments, and civil society groups all need to have a seat at the table. By engaging diverse perspectives, we can develop AI standards that are practical, relevant, and accepted by the broader community. This inclusive approach is key to creating governance frameworks that are both effective and sustainable.

Balancing flexibility with safety is one of the biggest governance challenges in AI. On the one hand, we want to encourage innovation and unlock AI's potential across industries. On the other hand, we must ensure robust safety measures are in place to prevent harm and maintain public trust. Striking this balance requires thoughtful governance models that evolve alongside the technology.

To conclude, the future of AI governance will be shaped by our ability to navigate complex challenges and foster collaboration. Whether it's managing autonomous systems, promoting cross-border cooperation, addressing ethical concerns, or adapting our regulatory approaches, a shared commitment to responsible AI development is essential. By working together, we can harness AI's potential for good while safeguarding society against its risks.

www.ingramcontent.com/pod-product-compliance
Lightning Source LLC
Chambersburg PA
CBHW060640210326
41520CB00010B/1686